U.S. Department of Transportation
Federal Highway Administration

Improving Traffic Signal Management and Operations: A Basic Service Model

Prepared by:

Booz | Allen | Hamilton

Iteris

Prepared for:

Federal Highway Administration

Washington, D.C.

DECEMBER 2009

FHWA-HOP-09-055

Notice and Quality Assurance Statement

NOTICE

This document is disseminated under the sponsorship of the U.S. Department of Transportation in the interest of information exchange. The U.S. Government assumes no liability for the use of the information contained in this document. This report does not constitute a standard, specification, or regulation. The U.S. Government does not endorse products or manufacturers. Trademarks or manufacturers' names appear in this report only because they are considered essential to the objective of the document.

QUALITY ASSURANCE STATEMENT

The Federal Highway Administration (FHWA) provides high-quality information to serve Government, industry, and the public in a manner that promotes public understanding. Standards and policies are used to ensure and maximize the quality, objectivity, utility, and integrity of its information. FHWA periodically reviews quality issues and adjusts its programs and processes to ensure continuous quality improvement.

Technical Report Documentation Page

1. Report No. FHWA-HOP-09-055	2. Government Accession No.	3. Recipient's Catalog No.	
4. Title and Subtitle Improving Traffic Signal Management and Operations: A Basic Service Model		5. Report Date: December 2009	
		6. Performing Organization Code	
7. Author(s) Richard W. Denney, Jr. PE, Iteris		8. Performing Organization Report No. Project	
9. Performing Organization Name and Address Booz | Allen | Hamilton 8283 Greensboro Drive McLean, Virginia 22102 and Iteris, Inc. 107 Carpenter Dr. Ste 230 Sterling, VA 20164		10. Work Unit No. (TRAIS)	
		11. Contract or Grant No. Contract No. DTFH61-06-D-00006	
12. Sponsoring Agency Name and Address U.S. Department of Transportation Federal Highway Administration Office of Operations 1200 New Jersey Ave, SE Washington, DC 20590		13. Type of Report and Period Covered Final Report December 2008 – December 2009	
		14. Sponsoring Agency Code HOP	
15. Supplementary Notes Eddie Curtis, Contracting Officer's Task Manager, FHWA. The project team acknowledges, Paul Olson, FHWA and Edward Fok, FHWA for their contributions to this report.			
16. Abstract This report provides a guide for achieving a basic service model for traffic signal management and operations. The basic service model is based on simply stated and defensible operational objectives that consider the staffing level, expertise and priorities of the responsible agency. The report includes a Literature Review, which provides a review of the National Traffic Signal Report Card and Self-Assessment, case studies based on agency archetypes that provide an understanding of how agencies deliver traffic signal management services based on their resources and interviews with acknowledged leaders providing support for the basic service concept. A discussion of signal timing versatility in support of the role it plays in providing good basic service is included in the Appendix.			
17. Key Words Signalized Intersections, Traffic Signal Timing, Staffing, Traffic Signal Operations, Maintenance, Management, Objectives, Performance Measures		18. Distribution Statement No Restrictions.	
19. Security Classification. Unclassified	20. Security Classification. (of this page) Unclassified	21. No. of Pages 47	22. Price N/A

Form DOT F 1700.7 (8-72) Reproduction of completed page authorized

Table of Contents

Notice and Quality Assurance Statement

Technical Report Documentation Page

Acknowledgments

About this Report

I. Background

II. Literature Review

    GAO Report to Congress and the ITE *Toolbox*
    Traffic Signal Report Card and Self Assessment

III. Case Studies

  Agency Archetypes
    Archetype 1. High-Activity Response to Adequate (Staffing) Resources
    Archetype 2. Infrastructure-Rich Response to Abundant (Capital) Resources
    Archetype 3. Well-Managed Response to Limited (Capital & Staffing) Resources
    Archetype 4. Insufficiently Managed Response to Limited Resources
    Archetype 5. Special Issues with Highly Dispersed Infrastructure (Typical State Agency)
    Features of Successful Archetypes
  Basic Service Concept
    Field Infrastructure Reliability
    Minimizing and Balancing Congestion
    Smooth Flow
    Predictable and Consistent Response
    Signal Timing Versatility

IV. Interviews

  Local Agency Traffic Signal Program Manager
  State Agency Traffic Signal Program Manager
  Observations

V. Implementing a Basic Service Concept: The Traffic Signal Management Plan

  Clarity of Objectives
  Attainable Performance Measures and Standards of Performance
  Field Infrastructure Reliability.
  Signal Timing
  Objective-Based Resource Allocation
  Clear Communication Up the Line
  Meaningful Systems Engineering
  Sample Outline for a Traffic Signal Management Plan
    Chapter 1. Objectives and Requirements.
    Chapter 2. Responsiveness to Citizens, Media, Policy Makers, and Elected Officials
    Chapter 3. Maintenance Strategies to Achieve Objectives
    Chapter 4. Operations Strategies to Achieve Objectives
    Chapter 5. Design Strategies to Achieve Objectives

Appendix 1. Signal Timing Versatility Concepts

Acknowledgments

This work reports on interviews conducted with experts who spoke from experience managing large and successful agencies. In order to provide the most candid responses, we have agreed not to identify them in this report, but we appreciate their wisdom and candor, and their willingness to discuss their day-to-day reality with the researchers.

The work also benefited from the expert review of FHWA subject-matter experts, including Paul Olson and Ed Fok (in addition to Eddie Curtis). The researchers appreciate their integrity in advocating for and tirelessly promoting good traffic signal management within agencies.

About this Report

Over the last several decades, the gap between the state of the art and the state of the practice in traffic signal operation has grown. Influential studies such as the 1994 General Accounting Office report to Congress have reported improved technological capabilities that go unused in agencies that increasingly find it difficult to meet the expectations of policy makers and the public.

Managing the expectations of the agency managers and decision makers to coincide with the purpose and objective of traffic signal operations has become an increasingly more difficult task for experienced practitioners and traffic signal system managers. The lack of clearly stated operational objectives may contribute to an unrealistic expectation that traffic signals can create capacity, when in fact they can only distribute it. Yet signal operations activities such as retiming projects and traffic signal infrastructure projects are most often proposed by those same practitioners to meet the expectation of relieving congestion.

The perceived gap between art and practice has led to a number of assessment efforts. In addition to the industry councils that led to such documents as the Institute of Transportation Engineers (ITE) document *A Toolbox for Alleviating Traffic Congestion and Enhancing Mobility*, and external assessments such as the GAO report, more recent efforts have included the Traffic Signal Report Card and Self Assessment. All of these efforts reinforce the assumption that activities are the measure of effectiveness. Thus, agencies seek to demonstrate that they are doing all they can to alleviate congestion by undertaking prescribed activities, including projects to retime traffic signals and projects to enhance the traffic signal infrastructure. These activities often consume more resources than are available to the agency, with the result that they dominate the agency's attention and priorities. By promising results based on activities alone, agency professionals often undermine their own credibility by setting unattainable expectations and then spending resources in competition with more attainable objectives.

The industry often promises what it cannot deliver, and then fails to deliver what it could, with better commitment and resources.

In this report, the research team has adopted a different approach. By identifying archetype agencies and then gathering information from real agencies, the researchers have attempted to define what makes an agency more successful with or without an abundance of resources. The report suggests that more successful agencies are those that establish and maintain a concept of *good basic service*. This concept becomes the basis for how they interact with the public, how they promise results to policy makers, and how they set priorities for limited resources. It focuses on results rather than activities.

The report begins with a background discussion and a review of the literature, particularly past assessments and focusing on the Traffic Signal Report Card.

Following the literature review, the report develops case studies based on archetype agencies. These archetypes illustrate the different levels of resources and how those resources are used, whether the agency looks good in assessments and whether it gets results in terms of motorist and policy-maker expectations. From these archetypes emerges the concept of good basic service. Targeted interviews with two agency managers provide examples of how well-regarded agencies define and achieve their goals (or not).

The interviews and the archetypes then form the basis for developing a basic service concept, including the outline for a Traffic Signal Management Plan for traffic signal operating agencies consistent with that basic service concept.

The Appendix includes a discussion of signal timing versatility and how the objective of versatility contrasts with the more common objective of optimality.

I. Background

Through the 1950's and 1960's, the technology and novelty of automatically controlled traffic signals led the agencies that owned and operated them to develop extensive staff capabilities to support their effective operation. Considerable resources were devoted to the development of guidelines and underlying research, leading to the common result that agency professionals held a high level of expertise and esteem within their agencies.

As traffic signals have become ubiquitous, their operation and management has emerged from a novel program receiving special emphasis to what appears to be a commodity. Agency professionals are less likely to be nationally known experts, less likely to be sent to national conferences where they can learn from their peers, less likely to be supported with society memberships such as in the Institute of Transportation Engineers, and less likely to have direct access to and the unconditional respect of policy makers and elected officials. This trend has led to a feeling of embattlement on the part of highly competent agency professionals, who are asked and sometimes demanded to do more with less as a matter of routine. The frequently seen turnover of elected and policy officials masks the effects of increasing demands even as such demands lead to a breakdown of agency effectiveness. This decline in agency regard mirrors the decline in other infrastructure-support elements within state and local governments.

Professional groups, both organized and ad hoc, have discussed these issues at length, and many methods have been proposed and attempted to bring greater public emphasis to the problem of declining support for traffic signal operation. In all cases, these efforts are hampered by the inability to easily measure the performance of the signal operation, lack of clarity as to the objective of the signal operation, and a simple understanding of how to achieve those objectives. Without clear goals and an equally clear plan to achieve those goals, agency professionals find it more and more difficult to articulate their resource needs within their agencies; even those needs critical to providing good basic service, let alone advanced capabilities. The various facets of this generalized characterization are vast. And despite that the characterization is a generalization, all agency professionals feel the pressure of trying to maintain good basic service in light of declining resources and credibility.

These trends have coupled with increasing urban populations and resulting increases in traffic congestion, which has undermined faith in the ability of traffic engineers to achieve real improvements. Agency professionals are often left with a losing proposition—to try to minimize the rate at which traffic conditions worsen, both as a result of increased demand and as a result of reduced field equipment maintenance.

The FHWA has the difficult task of supporting state and local agencies in their attempts to improve their ability to achieve better results. The reasons for this difficulty can be summarized as follows:

1. There is no consistent and defensible standard of good operation.

2. Measuring performance to test against a standard often requires unattainable resources.

3. As resources have declined, so have the ability of agencies to attract, develop, and retain true experts.

4. Many agencies depend on consultants to fill their expertise gap, but the turnover of consultants available to do such work is at least as high as the turnover within the agencies. This undermines attempts to improve the competence of practitioners through federally sponsored training, for example. The pool of consultant engineers with no agency experience continues to grow, in many cases widening the expertise gap.

5. Without a definition of good operation, there is no standard of good maintenance that is generally enforced.

6. Clear objectives and plans to achieve those objectives are not technical problems, yet the research community often focuses on technical issues, sometimes at levels far beyond the daily reality of practitioners. Research often focuses on technology to provide the solution, with many researchers believing that the hopelessness of developing proper standards and resources requires that meeting such standards will depend on expertise built into automated systems.

7. Practitioners often have difficulty providing agency policy makers and elected officials with a clear view of progress towards operational goals, and therefore are leaving them with a reluctance to fund improvements. Often practitioners who should provide this clarity feed the problem by not effectively managing expectations or by over-promising results in support of their projects. The classic example is the defense of traffic signal systems on the basis of performance measures wholly related to improved timings, without the recognition that improved timings can be implemented without the system. Thus, systems are not justified by the true needs, which are based on managing field infrastructure as opposed to attaining specific traffic performance improvements.

The FHWA has promoted a number of efforts designed to measure and improve the situation. The Traffic Signal Report Card, which was conducted by the industry in support of a major FHWA initiative, showed generally poor results around the country. But those results were often based on effecting positive outcomes by completing a variety of activities, or by maintaining arbitrary ratios such as the frequency with which signals are retimed or the number of maintenance technicians on staff.

FHWA also has supported the development of a range of training programs, including National Highway Institute courses on computerized traffic signal systems and on traffic signal design and operation, and on Internet-based courses such as on traffic signal timing. These courses help, but they must be part of an overall program to evaluate, reward, and maintain the professional capacity of agency staff. Often, the students of these courses have little authority to implement the recommendations offered, and the policy makers who do have that authority are rarely involved.

More recently the FHWA has developed a Traffic Signal Timing Manual and is currently promoting the formation of Regional Traffic Signal Management Programs. Regional collaboration allows signal operators to leverage the expertise and resources within the region to clearly articulate operations objectives and performance measures that resonate at a region level often attracting support for and investment in the management and operation of signal systems.

Making progress in the context of these very difficult issues is a tremendous challenge, both technical and organizational. Organizational improvements must support technical improvements, and technical improvements are necessary to meet organizational objectives. Instilling those objectives and improvements will require achieving a clarity of understanding concerning where traffic signal management should be hopeful and where expectations should be moderated.

The industry often promises what it cannot deliver, and then fails to deliver what it could, with better commitment and resources.

II. Literature Review
GAO Report to Congress and the ITE *Toolbox*

In March, 1994, the General Accounting Office published a report to the Chairman of the House of Representatives Energy and Commerce Committee. That report was titled, *Transportation Infrastructure, Benefits of Traffic Control Signal Systems Are Not Being Fully Realized.* That report was motivated in part by a 1990 review conducted by the Federal Highway Administration that found that 21 of 24 traffic signal systems did not meet minimum standards of performance. That assessment, titled *Operation and Maintenance of Traffic Control Systems,* was motivated by the Institute of Transportation Engineers publication *A Toolbox for Alleviating Traffic Congestion and Enhancing Mobility,* which had documented that many local agencies were installing systems without considering the cost of proper operations and maintenance. In 1992, a panel was convened to review the FHWA Report and make recommendations for improvements. These recommendations led to the establishment of federal funding for operations and maintenance, which was included in the ISTEA legislation.

The following bullets are some of the recommendations that emerged from the ITE panel:

- Rules and procedures for Federal funding of operations and maintenance.
- Minimum standards for the knowledge, skills, and abilities required for operations and maintenance
- Guidelines for staffing
- Model plans for operations and maintenance
- New training courses covering operations and maintenance
- Formation of regional traffic management committees.
- Require state and local agencies to develop staffing plans
- Require state and local agencies to lead regional traffic management committees

The additional recommendations (there were 34) were related to traffic signal systems, which was the focus of all these efforts.

The GAO report in 1994 summarized the resulting activities and improvements. The report described external issues that contributed to the problem, the main one being jurisdictional cooperation. It also described the emerging ITS technologies and how they should be incorporated, and it described FHWA's activities in responding to the above recommendations.

The main weakness of the report is that it did not describe good operation, or establish a meaning for good basic service. As such, most local agencies still faced the constraints that held them back previously. The Congestion Mitigation and Air Quality Program in the ISTEA funding legislation did, however, provide significant new funding for traffic signal systems, and ISTEA did provide funding for operations and maintenance for some state and local agencies.

The conclusion that the capabilities of new signal systems were being wasted was made without clearly defining what the needs were for basic operation, and thus without knowing whether those features were actually useful in as many situations as the report suggested. Also, the report did not determine whether these advanced features really added to the cost of those systems. The report was correct in noting (and in summarizing the results of those previous studies) that agencies were using capital funds to buy technology rather than focusing on good basic service.

In 1997, ITE updated the *Toolbox,* with considerable expansion to cover emerging ITS activities. For traffic signals, the *Toolbox* included signal timing improvements, updated equipment, coordination, removal of unneeded signals, and improved maintenance as tools in the toolbox. They provide an estimate of the benefits and costs of these tools, but the report did not attempt to define good basic service, or to provide examples or information of how agencies to define their own basic service concept.

Traffic Signal Report Card and Self Assessment

In 2005 (and then updated in 2007), the ITE under FHWA sponsorship developed a national report card identifying the overall effectiveness of signal operations and management. The report card was based on a self-assessment process, which itself identifies a range of priorities and objectives for agencies by grading agencies on those objectives.

The contents of the self-assessment include the following categories:

- Management
- Intersection signal operation
- Coordinated system operation
- Signal timing practices
- Traffic monitoring and data collection
- Maintenance

Each of these categories included general questions on overall programming, but more importantly for this review are those questions that reflect specific objectives. The following sections discuss each category, with an attempt to determine the underlying objectives that might be used to guide traffic signal management.

Management

Activities considered important enough to ask about included:

- Commitment of staff to peak period active traffic management
- Motorist information
- Special pedestrian needs
- Automated active monitoring of traffic flow
- Ability to respond to planned events
- Ability to respond to unplanned events

All of these objectives are based on traffic conditions and control responses. Good management was described as including active programs for tracking events, monitoring incidents, and so on. None of the questions, however, determined if the agency had an operational objective, other than being responsive. The self-assessment, however, does not query agencies concerning the importance of these activities in their jurisdiction other than a general question of whether the agency had an overall operational philosophy.

Intersection Signal Timing

Important activities included:

- Process documentation in place for routine retiming
- Signal timing database management
- Routine reviews should include pedestrian volumes, vehicle volumes, and speeds
- Are clearance intervals and other non-operational timings reviewed routinely?
- Follow-through of reviews to implementation and fine-tuning
- Review other control aspects, including the need for the signal, phases, and restrictions.
- Review railroad operation

As with the Management category, there is no discussion of whether agencies have an operational objective. It does not question whether agencies have a consistent clearance-interval policy, a consistent pedestrian accommodation policy, etc. It does not make the distinction between design signal timing parameters and operational parameters. Design parameters include clearance intervals, detector timing, minimum greens, and pedestrian intervals which are based on the geometry of the intersection. Assuming a policy was in place for calculating these parameters, well-managed agencies might not see the importance of reviewing them routinely, except to determine that the intersection geometry has not changed, and the signal timings in the controller are the intended settings (which *is* covered by the database management question).

Coordinated Operation

Underlying measures of effectiveness in this category include:

- Reviewing coordination timing at least every three years.
- Reviewing coordination parameters more often as needed.
- Timely follow-through following review
- Optimization software is a requirement of routine retiming
- Accommodating weekends, holidays, and planned events
- A policy for when adjacent intersections should be coordinated
- Coordination should cross jurisdictional boundaries
- Traffic-responsive or adaptive control is considered the correct response to unpredictable demand

Again, no attempt is made to determine whether these measures really are relevant to strong operational objectives. For example, if the objective is to minimize travel time, or, in the absence of that, to minimize inequity, then traffic-responsive operation might not be the solution even if traffic volumes are not completely predictable. Rather, an agency may properly respond to this objective by purposely designing signal timings for maximum versatility. A more fruitful approach might be to determine if the agency has identified, implemented, and evaluated a strategy for responding to situations of unpredictable demand. Given that the concepts of signal timing versatility are not well established in the practice, questions may need to be tailored to specific approaches that promote versatility, such as bandwidth optimization in response to unpredictable demand, which a practitioner might prefer because it addresses a wider range of conditions than calculated optimal timings based on a specific, and possibly not representative, set of conditions.

Signal Timing Practices

Underlying objectives in this category are:

- Including queue-departure time in progression timing
- Consider all parameters using software
- Willingness to vary phase sequence to enhance progression
- Minimize delay and stops during night-time operations
- Avoiding turn-lane spillover

These objectives are the most closely related to the general objective of minimizing travel time and providing equitable operation. Several, however, reflect regional practices that are not applied everywhere. For instance, before an agency implements variable phase sequence, which may be an unpopular or unexpected new operation in that region, agency practitioners should determine whether it actually promotes their operational objectives. A key feature in these objectives is the implicit assumption that computer optimization is an important part of the process. This implies a deep requirement for data collection and management that serves no other purpose than feeding the software, and may not be tied directly to agency objectives. Thus, the use of practices such as variable phase sequence does not mark an effective agency as well as the agency's process of considering whether such practices meet their objectives.

Traffic Monitoring and Data Collection

This category implies the objectives of:

- Collecting, evaluating, maintaining and sharing traffic data
- Measuring system performance
- Reporting activity to decision makers, elected officials and the public

The traffic data collected including turning movement counts in 15-minute intervals, and performance data including travel time and number of stops. Again, the data are not directly tied to objectives to determine the true importance of the data.

Maintenance

The underlying objectives implied in this category include:

- Responding to emergency maintenance within 1 hour during business hours.
- Tracking funding with growth, life-cycle replacements, and technological obsolescence.

- Implementing a process to receive, track and respond to citizen complaints.
- Maintaining high levels of technician training and certification.
- Reviewing and routine testing of signal equipment frequently (quarterly through annually, depending on the review or test).
- Automated malfunction monitoring.
- Maintaining high detection functionality.
- Tracking equipment maintenance trends.
- Providing emergency power at critical locations.

These objectives allow agencies to score well if they have well-funded and active maintenance programs. Not all agencies have those resources, however, and signal practitioners still must respond to their situation as best they can. There is nothing in the self-assessment that allows an agency to overcome limited resources by more effective design. For example, an agency might achieve more effectively meet their operational goals by minimizing the number of detectors and designing their operation in such a way that those detectors are not needed. This might be done by limiting the cycle length, providing versatile smooth-flow coordination timings, ensuring that clearance intervals are sufficient to minimize the need for dilemma-zone detection, and so on. Highly durable detection construction might be affordable if the detection needs are minimized, further reducing the demands on maintenance. Consistent design standards (rather than high design in some locations that are funded by developers and minimal design in locations funded by the agency) can minimize the need for continued maintenance of the physical plant. Careful selection of technology might prevent expecting maintenance technicians to have skills beyond what the agency can remunerate. All of these strategies for surviving limited maintenance resources flow from a careful relationship between operational objectives and maintenance realities, but the self-assessment does not credit agencies who exercise such care.

Summary

The Traffic Signal Report and Self-Assessment was successful at shining the spotlight on the need for greater investment in traffic signal operations and maintenance, and providing a benchmark of common practices within the traffic signal community. Effectively, the Traffic signal report card is a document for policy makers rather than for practitioners, but it does imply some global objectives and standards of performance for local agencies that may not always be consistent with the realities faced by practitioners. The Traffic Signal Audit Guide also developed by NTOC offers an opportunity for practitioners to take a more in depth look at their specific practices and to develop process that more effectively meet local objectives.

When agency professionals are developing and defending their budgets, the Report Card and its associated reports and audit guide are most helpful, and collectively these documents provide the only real synthesis of expected and good practice. In the end, though, it is a diagnostic tool at the screening level, not a clinical approach for a given agency to improve their traffic signal management.

III. Case Studies

Agency Archetypes

Archetypes of agencies can be used to evaluate the effectiveness of agency practices and needs. In this section, we will lay out a range of possible agency archetypes, and then we will use these archetypes as the basis for a series of interviews with real agencies.

Archetype 1. High-Activity Response to Adequate (Staffing) Resources

Archetype One would be typical of a state or large metropolitan agency that has identified operations and maintenance as a core business practice. The business plan for that agency supports staffing and resources focused on delivering an operations program. The staff undertakes bi-annual signal timing reviews by hiring outside consultants to collect turning-movement counts at all intersections, reviewing all fixed intervals at each intersection, developing and programming signal optimization software for calculating optimized signal timings, evaluating signal timings using micro-simulation, and providing revised signal timings to the agency. The agency has an active program for maintaining the data and making sure it is installed and maintained in signal controllers and systems.

Before any signal timings can be implemented, they must be evaluated using microscopic simulation, both by consultant and agency staff. Agency staff are hired with extensive expertise in analysis and simulation, and encouraged to improve that expertise through regular training. Consultants are hired to work on site, under agency supervision and with no access to highly experienced consultants. The operations staff does not have the authority over basic timing parameters such as phasing, phase sequence, pedestrian accommodation, detector placement, and intersection lane usage. Those parameters are aggressively controlled by other departments within the agency. Because of the perceived workload resulting from high activity levels, operational personnel also are not encouraged to spend time in the field to observe the results of the operation. Using microsimulation was a response to this restriction (and also a cause of it). Improvements are reported on the basis of simulation.

This archetype sets a precedent for for detailed analysis. However, several key weaknesses in their situation prevent getting the best results:

- Not being able to design phase sequence along with signal timing limits the versatility and optimality of the results. Often, the phase sequence decisions are made to protect as many movements as possible rather than to provide efficient operation.

- Signal timing technicians and engineers do not have authority over final results, and are not encouraged to understand their analysis in terms of on-street performance.

- Traffic engineering staff and operations staff are not mutually responsive, and some operation is controlled by one group while other operation is controlled by a separate group. This is quite common in real agencies, where a traffic signal system is operated by a different group than the basic signal timing staff that does non-systems operational design including phase sequence.

- Befitting their abundant resources, the agency employs or contracts with a large pool of engineers. Those engineers, however, remain separated from maintenance technicians. The technicians and traffic engineers do not benefit from any expertise in operations, and the operations staff often work in the absence of relevant input from traffic engineers and technicians.

- The timing approach is driven by software and the data required by software, rather than being driven by performance results on the street.

- Most of the agency's considerable operations resources are consumed doing data collection and analysis, rather than doing on-the-ground performance measurement meaningful to the objectives of motorists.

- The agency does not benefit from hiring consultants, using them as an extension of staff rather than a source of expertise. This is reflected by structuring their consulting projects to require on-site services, which makes it difficult for consulting companies to commit their most experienced experts, and by refusing the pay the rates of consultants who are most advanced in their careers. (This report does not attempt to address the problem that consultants are not consistently expert despite their rates or stated experience.)

- The resources and software emphasis reduces the agency's ability to respond to special-event signal timing needs.

- The agency responds to citizen complaints with descriptions of activities they are undertaking, rather than with the reasons for the operation being complained about. Complaints lead to studies, justifications for system enhancements, revised timings, and other activities, but operations staff are not compelled to be able to explain their decisions to citizens and their representatives.

- The agency believes that it is undertaking an aggressive and successful program.

Archetype 2. Infrastructure-Rich Response to Abundant (Capital) Resources

Archetype Two might be typical of an agency that has committed to improving operations through capital programs that supports infrastructure improvements. The agency has not garnered support for increasing staffing resources to support active operations; which, effectively is a new activity within the agency. The agency has built an extensive centralized traffic signal system with downstream detectors at all intersections, adaptively controlled. Operations staff is limited but maintenance capabilities are good and capital programs well-funded. The limited operations staff are used to performing travel time analysis and systematically review the network for unexpected queuing. The agency has a defined operational mission of providing smooth flow and minimizing congestion, the technicians are trained to observe and evaluate the system in accordance with those objectives. The strong maintenance capability keeps the extensive infrastructure in good working order. Well-founded design standards provide consistent infrastructure (that provides the features needed to achieve their operational objectives) and pedestrian accommodation, but without much downstream review after construction.

The agency reviews signal timings yearly by scheduling and reviewing split monitoring in the signal controllers, for each arterial system. This allows them to identify non-coordinated movements that consistently max out, suggesting the possibility of residual queuing. This is accomplished using the two operations technicians on staff. The two staff members work a split shift, with one technician working during the morning rush and the other during the afternoon rush. The morning technician observes system data for an arterial system Tuesday through Friday mornings, and the afternoon technician collects similar data Monday through Thursday afternoons. The technicians handle adaptive parameter and new intersection data maintenance duties in the middle of the day, when their shifts overlap. Using this approach, they are able to systematically review all 175 arterial systems plus a downtown grid each year. The split monitor and adaptive data results are summarized and reviewed by the engineer in charge of the signal program during the same week. Unexpected queuing or anomalies lead the engineer to schedule a site visit and a review of the arterial. If a field evaluation cannot rectify the problem, a more detailed analysis is conducted using data collection and analysis.

The focus of agency staff on actual performance observation allows them to be responsive to citizen complaints. Complaints are handled by the engineer in charge with input from the technicians who review that location. The signal system provides real-time monitoring capabilities that allow operations staff to respond substantively to citizen complaints that are caused by malfunction. The review technicians are also efficiently used for actual observation and assessment rather than for activity-based strategies, such as data collection and analysis, which may not always serve their objectives efficiently.

This agency does not score well on the Report Card because of limited systematic data collection and retiming and management that is not focused on real-time human interaction. But it shows strong results because:

- Insufficient operations staff resources are compensated by effective capital expenditure on systems that work effectively for long periods without frequent systematic retiming.
- The agency is focused on performance measures appropriate to motorist objectives, as a validation of the ongoing adaptive control.
- By including pedestrian needs systematically in the design process, the need to be responsive to pedestrian requests is reduced down to special cases.
- The agency designed a system that made use of their strengths (strong maintenance) and minimized their weaknesses (limited operations staff).

Archetype 3. Well-Managed Response to Limited (Capital & Staffing) Resources

Agency Three has limited operations and maintenance staffing and limited capital resources. The operations staff is small, but the agency has a well-developed operational objective of providing free-flow travel time or equitable operation when that is not possible, as stated in a mission:

Don't make motorists stop, but if you must, delay them as little as possible.

The agency has difficulty maintaining extensive detection and does not build intersection detectors in coordinated systems except on minor-street approaches. The agency also focuses on efficient and versatile timing that will operate reasonably even when detectors are not working. For example, they limit controller max times based on engineered fixed time operation, and they keep cycles short and use the least number of protected phases possible. Design favors progression for the predominate through flows. Operations staff are limited but have authority over the full operation of the signal, including design and signal phasing.

The agency reviews signal timings yearly using floating-car studies on each arterial system. This is accomplished using the two operations technicians on staff. The two staff members work a split shift, with one technician working during the morning rush and the other during the afternoon rush. The morning technician collects travel time data on an arterial system Tuesday through Friday mornings, and the afternoon technician collects similar data Monday through Thursday afternoons. Five runs are made in each direction. In addition to collecting travel time data, the technicians review operation by having a time-space diagram available in the vehicle to ensure that the signals are operating as designed. The technicians handle timing database maintenance duties in the middle of the day, when their shifts overlap. Using this approach, they are able to systematically review all 175 arterial systems plus a downtown grid each year. The travel time studies are compiled during the middle of the day, and reviewed by the engineer in charge of the signal program during the same week. Unexpected queuing or other delay noted in the travel time studies lead the engineer to schedule a site visit and a review of the time-space diagram. If a field evaluation cannot rectify the problem, a more detailed analysis is conducted using data collection and Synchro optimization by consultants. Technicians require training in travel time data collection and in observing that signal timing matches time-space diagrams, but it does not require expertise in simulation or optimization modeling.

The agency also built signal systems that provide remote access to local controllers to minimize the need for field visits to respond to complaints, but without traffic-responsive or adaptive control, which requires infrastructure (both communications and detection) that would have prevented agency-wide system implementation. The agency focuses its limited design resources on setting clear standards for good design such that pedestrian accommodations, phasing, initial timing, and lane utilization are done right at the outset. For example, during design, a complete operational review is performed to make sure that each signal phase provided is needed, that lane utilization balances capacity with respect to demand, and that detection (both pedestrian and vehicular) critical to minimizing the impact of the signal on major movements is provided. The design principles favor detection where it serves the objective of not stopping cars on the main street, without, for example, the added cost of extensive detection on the main street. (A common alternative to these appropriate standards is the use of typical designs without evaluating their needs at a given intersections, including but not limited to always providing a left turn phases when left turn lanes are provided, not considering whether a through lane should really be a turn lane, automatic use of a specific phasing scheme—such as 8-phase operation with leading lefts—without considering the demand at the intersection, providing extensive dilemma-zone detection even on intersections in coordinated systems, etc.) Operational staff are limited but have full responsibility and authority for the operation of the intersection, including phasing, phase sequence, signal head arrangements, detection, and timing.

While Agency Three's funding is considerably less than Agency One's, they get better results, and further they can show that their results are better, because:

- The agency devotes its limited resources to evaluating the operation based on a clear understanding of motorist objectives. The objective of not stopping traffic, for example, is directly measured by the low-cost travel-time studies, rather than the costlier collection of traffic volumes and attempts to model, based on those volumes, a range of performance measures that have not been clearly articulated and are therefore difficult to evaluate.

- The agency developed and then consistently applies design standards consistent with a realistic assessment of their maintenance resources. Thus, they don't end up with some intersections designed at a high level and others that don't meet adequate standards, particularly in regards to detection.

- The agency designed signal systems consistent with their maintenance and operations capability. Their strong, objective-based measurement and evaluation program was considered more useful in responding to motorist needs than in providing additional automatic traffic monitoring. Systems with more advanced detection and control would have exceeded their budget for city-wide installation, and they considered it better to have good field device management throughout the agency rather than advanced features in a smaller area and no capability elsewhere.

- The agency is highly responsive to citizen input because of their motorist-relevant measurement approach and good access to field infrastructure.

- The versatility of signal timings minimizes the need for additional detection and automated responsiveness.

- Centralized authority over all aspects of design and operation provide the tools necessary for versatile signal timing.

- Consultants are used effectively for situations where new timings are truly needed.

This agency might not score as well as Agency One on the Report Card, but they get better results with respect to their more limited or constrained resources. Those better results are reflected in operations consistent with their operational objective—rather than engaging in office and computer-based data collection and retiming efforts, they are on the street observing actual operation from the perspective of the motorists. The time spent by Agency One staff developing simulation models and other volume-based analysis is spent by Agency Three to evaluate actual operation in the field, based on measuring travel times directly. And flags for action are based on those travel time measurements, which will quickly reveal new congestion, rather than on an assumption of need implied by an arbitrary schedule.

Archetype 4. Insufficiently Managed Response to Limited Resources

This agency has the same resource limitations as in the previous scenario, but strives implement a program of activities without clearly articulating the objectives and outcomes they would like to achieve. These activities are accepted as common practice but without clearly mapping the activities to local objectives and performance measures the outcomes may not be improved. This agency might perform well on the Report Card but not experience the desired level of on-street performance for the traffic signal program. The agency opts to hire consultants allowing them to apply arbitrary standards for design and operation rather than focusing on their limited resources, establishing consistent and adequate design standards and operational objectives. The outcome of this approach is expensive, detector-rich installations, resulting in fewer intersections being built due to the high costs of construction and maintenance. This forces the agency to build some intersections at extremely low cost that do not comply with any principles of good design. Thus, some intersections are built with an abundance of detection, mast arms, and expensive signal head arrangements, while others are built on wood poles using span wire, the minimum number of heads, and no pedestrian accommodation. This leads to intersections that must operate poorly to satisfy MUTCD pedestrian accommodation requirements.

Signal timing is performed by consultants without clear objectives, resulting in timings that attempt to make the most out of the provided infrastructure. Thus, intersections with extensive approach detection use dilemma-zone schemes and long actuated cycles that depend on actuation to provide optimal operation. Max times are set long to allow actuation to work. Consultants based their design on data collected by the agency, because of limited budgets to pay consultants to do so. The two operations technicians on staff are used to set road-tube counts and conduct occasional turning-movement counts. Signal timings are developed for all arterials on a rotating, five-year program. To improve the effectiveness of signal timing, intersection detectors are tasked as system detectors, and arterial systems use traffic-responsive operation. Frequent detector failures greatly reduce the effectiveness of the operation, which falls back to programmed daily schedules, but without the agency's frequent review.

The large number of detectors at some intersections pose a maintenance challenge, and the agency cannot keep the detectors working. Thus, the intersections do not operate effectively, with detector failures resulting in long green times and cycles even when demand is low. Limited resources preclude the needed reviews of these intersections.

The only review mechanism is responding to citizen complaints. When a complaint is received, the agency attempts to review operation on the basis of detection, even if several intersections away. The engineer reviewing each complaint has no baseline travel-time-based performance data against which to compare current operation. Signal timings are tweaked to accommodate the complaint, but without a systematic review of the whole arterial, which must wait for the next signal timing update, those tweaks may undermine the basic versatility of the signal timing plan. For example, adjustments to improve intersection splits often undermine progression if the progression design isn't considered as part of the adjustment.

The results achieved by the agency are poor, because:

- The agency has no performance measurement outside of citizen complaints, let alone measurement linked to motorist objectives.

- The agency feeds analysis software rather than collecting performance data.

- The agency is reactive rather than proactive.

- Adjustments made between optimization projects are not made systematically within the context of the timing design.

- The operational picture presented to motorists is inconsistent, with apparent inequities.

- Operation is usually degraded by the failure of detection, mostly because the operation depends too much on the detection.

Because the agency collects data and performs analysis, it might score better on the Report Card than in the previous scenario, though the results will not be as good, because resources are not aimed at activities directly linked to performance that matters most to motorists. Those activities can be seen as one step removed from operation on the street, as an abstraction or representation of that operation, rather than as the definite example. Analysis might show a significant improvement in delay, for example, but motorists are more directly concerned with how smoothly and predictably they flow through the network. The temptation to depend on analysis more than observation overwhelms this agency.

Archetype 5. Special Issues with Highly Dispersed Infrastructure (Typical State Agency)

Archetype Five is typical of an agency that has limited operations and maintenance capabilities due to in adequate capita and staffing resources relative to the number and dispersion of traffic signals within its jurisdiction.

Most of the intersections built by the agency are in remote areas, and coordinated systems are on suburban or fringe arterial streets and not grid networks. The arterial networks frequently center or cluster around freeway interchanges, including diamond interchanges.

The key operational constraint on the agency is the distance from the office to the traffic signals. Often, a state DOT district office is located in a city that maintains its own traffic signals. And just as often, the agency's operations staff must divide their time with many other traffic engineering duties, including signs, pavement markings, and liaison with local agencies.

Because of this, these agencies usually seek to make their intersections operate without their direct involvement to the extent possible. These limitations lead to a series of broad requirements:

- The operation of the signal must accommodate changes in traffic at all time series, including daily, weekly, and seasonal variation in addition to long-term trends.

- The signal must provide access from the operations staff for observing operation in response to citizen complaints, and for making adjustments and maintaining the signal timing database. This access is needed on an ad hoc basis.

- Field reliability is paramount. This requirement is in conflict with the first requirement, because accommodating traffic variation over the long run requires detection, and detection is the least reliable component of traffic signal installations.

- Design standards that focus on high-quality installations improve reliability and thus minimize the need for future involvement. This includes not only detection, but also the physical infrastructure of the signal to support a range of operation (such as using longer mast arms than necessary to ease the installation of future left-turn heads). An agency with more signals closer to their office might be able to adopt a model that requires greater on-street involvement.

The agency reviews signal timings yearly using floating-car studies on each arterial system. Unexpected queuing or other delay noted in the travel time studies lead the engineer to schedule a site visit and a review of the time-space diagram. If a field evaluation cannot rectify the problem, a more detailed analysis is conducted using data collection and Synchro optimization by consultants. Signal timing is designed expressly to provide the maximum potential for through-band progression. Given the fringe suburban setting of their systems, severe congestion is rare, but rapid growth means that variability is high and such versatility maintains acceptable operation in the face of such variability.

The agency also built signal systems that provide remote access to local controllers to minimize the need for field visits to respond to complaints, but without traffic-responsive or adaptive control, which requires infrastructure (both communications and detection) inconsistent with the long distances to the office. The agency focuses its limited design resources on setting clear standards for good design such that pedestrian accommodations, phasing, initial timing, and lane utilization are done right at the outset. Operational staff are limited but have full responsibility and authority for the operation of the intersection, including phasing, phase sequence, signal head arrangements, detection, and timing.

Agency Five uses the detection infrastructure, and the ability of their systems to give them access to the field equipment to minimize the effect of such large distances to the signals from their office. Thus, they get good results:

- The agency devotes its limited resources to evaluating the operation based on a clear understanding of motorist objectives.

- The agency applies design standards consistent with a realistic assessment of their maintenance resources. For this agency, that means high-quality installations to minimize maintenance visits, which for them is a more important measure than maintenance skill.

- The agency designed signal systems consistent with their maintenance and operations capability. Their need to monitor conditions remotely, and to effectively manage the field infrastructure remotely is well accounted in their design.

- The agency is highly responsive to citizen input because of their motorist-relevant measurement approach and good access to field infrastructure.

- The versatility of signal timings minimizes the need for additional detection and automated responsiveness.

- Centralized authority over all aspects of design and operation provide the tools necessary for versatile signal timing.

- Consultants are used effectively for situations where new timings are truly needed.

Features of Successful Archetypes

The archetypes above that are successful present the following features:

- **Strong concept of basic service.** In the face of limited objectives, the effective agencies place a high priority on providing good basic service first, with less emphasis on attempting advanced service models. These agencies have defined what basic service means for their agency, and they evaluate their operation in terms of those basic services to ensure that they succeed. The ease with which one can construct an agency archetype that spends a lot of resources without achieving good basic service demonstrates that even good basic service can be a challenging goal. And while the Report Card does not distinguish between basic and advanced service, the poor Report Card performance suggests that focusing on basic service is not as common as it might be. Most agencies have limited resources, and all agencies experience occasional resource reductions; the definition of and commitment to good basic service is highly recommended.

- **Clear evaluation of objectives.** While minimizing travel time might be considered too broad by many practitioners, the fact that only Archetype Three responds to limited resources by focusing more tightly on that simple objective reveals too little attention to those broad objectives. This is closely related to the basic service concept, and the best agency models evaluate their effectiveness in terms of basic service objectives first. Archetype One conducted extremely detailed analysis, but without on-the-ground attention to simply stated broad objectives. Thus, it cannot really know how well it meets that objective.

- **Close coordination of design, operations, and maintenance resources and limitations.** The successful archetype agencies avoid constructing infrastructure elements that cannot be maintained, and they avoid operation that demands such infrastructure and maintenance. They also avoid overdesigning some intersections at the expense of barely meeting minimum standards at other intersections. Or, they purposely design systems that demand more capital funding up front in order to respond to a limited operational capability of their staff. The relationship between design, operations capability, and maintenance resources is the key element of this feature. Each agency must develop that relationship for themselves based on the dynamic balance between resources and objectives. Increasing infrastructure sophistication may be a good trade-off in response to limited operations staff. *Decreasing* infrastructure sophistication may be a good trade-off in response to limited maintenance capabilities. An agency should not build what it cannot maintain and operate, but sometimes a more sophisticated design may reduce demand on maintenance and operation. For example, one successful archetype used adaptive control to reduce the operations resource requirements, but with the recognition that adaptive control placed a greater demand on the agency's maintenance capability. An agency limited in both maintenance and operations may achieve the best balance by focusing on simple and highly versatile coordinated signal timings rather than on detection-intensive designs that result in poor operation when the detection fails. And the choice to use simple and versatile timings may get more consistently acceptable results rather than depending on analysis-intensive timing design which narrows the optimal range and reduces versatility in return for marginally better measures of effectiveness as reported by the analysis tools.

- **Good understanding of measuring results.** This is related to the strong concept of basic service, but it also leads to a better validation process for more advanced service. For example, an agency that is spread over a large geographic area (the limitation) may design systems that maximize remote monitoring capability (the response), on the basis of minimizing the number of required field visits (the measure). An agency that has limited staff resources but with high skill levels may more effectively invest in sophisticated automated analysis methods, while an agency with a larger staff but with lesser skills might need to keep their analysis methods simple, even though they require more effort to implement.

- **Commitment to staff development.** The strong archetypes seem to better understand the capabilities of their staff and do their best to develop staff capabilities and reward progress. It should be noted that not all rewards must be remunerative. Agencies that provide clarity of objectives and direct methods for evaluating achievement in those objectives often observe higher staff satisfaction and enjoyment of their work. And it should also be noted that staff development includes a mix of activities, including specific training to improve knowledge, mentoring to improve skills and abilities, and support for external professional activities such as attending conferences and being involved in regional and national technical, policy, and research committees.

Basic Service Concept

In this section, the concept of basic service, and how basic service might be evaluated, will be further developed.

As stated in the introduction, any basic service concept must start with the expectations of the motorists to whom the service is being provided. Most motorists, when driving their cars (as opposed to when they render opinions in the public forum) have some variation on the following broad expectation:

I want to drive to my destination at my desired speed with the minimum of attention and interruption. In the absence of achieving that goal, I want to be treated fairly and predictably so that I can plan my day with the minimum of uncertainty.

While such characterizations seem ambiguous, they do lead to a number of clear objectives for an agency. These might be:

- Field infrastructure reliability.

- Signal timing that minimizes and balances congestion.

- Signal timing that promotes smooth flow. This objective is an amalgamation of more narrowly cast objectives, such as delay, stops, and so on. Actions taken to achieve those narrow objectives should be evaluated in terms of broad objectives. Delay may be reduced numerically, but will the action result in smoother flow? This will usually require the application of experienced judgment, but not necessarily high levels of technical sophistication.

- Signal operation that responds to conditions predictably and consistently.
- Versatile signal timing that provides broad-banded solutions rather than being narrowly optimized for a specific condition.

Each of these agency objectives requires a series of fundamental agency requirements to support them. These requirements can form the basis for evaluating basic service.

Field Infrastructure Reliability

This objective imposes the following requirements:

- Realistic assessment of maintenance capability. Most agencies are resource-limited because maintenance is usually funded by a jurisdiction's general fund, rather than by capital funds, and those resources are capped and subject to sometimes fierce competition within the local policy and political environment. General fund sources include local property and (rarely) income tax, with some programs supporting operations and maintenance from the Federal government. Capital funds are supported by relatively rich Federally funded construction programs and the sale of bonds. This reality, which is usually outside the control of the traffic signal professionals, nevertheless requires those professionals to consider how they manage their resources. The two main constraints of maintenance resources are technological and quantitative.

    o Technological resource limitations affect the level of technology an agency can support on the basis of their salary structure, training environment, general work environment, and local labor pool. Of these facets of the problem, only the training environment can be modified by the agency on the basis of a policy decision at the level of the operational staff. Agencies therefore attempt to train their best technicians to handle more advanced technologies, but often then lose those technicians to more lucrative employment in often more pleasing work environments, especially if such skills are in demand in the local labor pool. An appropriate response to this limitation, other than articulating it in hopes of broader policy-level understanding and support, is to design infrastructure and adopt operational approaches within the technological capabilities of the agency's maintenance resources. Another approach might be to devise a maintenance concept that rents that capability from a contractor rather than trying to respond within the agency's forces.

    o Quantitative limitations abound in traffic signal agencies. Most agencies of any size complain that each maintenance technician is spread too thin over too many traffic signals, and this complaint is heard even by the most well-funded agencies. One approach to responding to this limitation is to reduce the complexity of the infrastructure. This is distinct from reducing the technical depth of that infrastructure, and focuses more on the number of infrastructure elements built. For example, the most vulnerable infrastructure element in any traffic signal program is detection. One of the main reasons for the rise in popularity of video detection systems, despite questions of their accuracy in certain situations, is that they are perceived to be easy to maintain, particularly without having to block traffic, and easy to reconfigure. A system that can provide effective signal timings with *less* detection can also achieve this objective.

- Designing within an agency's capabilities. In a forum on operations and maintenance held during the late 1980's, Anson Nordby, then the head of the City of Los Angeles ATSAC program, observed during a panel discussion that designers often treat each system project as an opportunity for a technological *tour-de-force.* He further observed that by doing so, the designers were often leaving the agency's maintenance capability behind. The wisdom in this observation suggests that one critical step in any signal and signal system design project is a careful evaluation of the agency's maintenance requirements. For example, signal designs that require bucket trucks that first-line maintenance technicians do not use will require failures of those systems to be responded to by second-line resources. This increases the time required to respond to a problem, which undermines the basic service objective. Detection again provides another example. Many agencies are unable to maintain their pavement as well as they would like, and poor pavement usually results in poor in-pavement detector longevity. It should be noted that the only way designers will understand maintenance limitations is to ask the maintenance staff.

- Using capital resources to minimize maintenance impact. One approach to the problem mentioned above of detector failures in poor pavements is to improve the design and construction of the detectors with the poor pavement considered. The City of San Antonio, in response to a large number of failed detectors, and in addition to moving to operational approaches that minimized the need for detectors, started constructing them more durably. They used a 4" rock saw and made a 6"-deep cut. The loop detector wiring was enclosed in Schedule 80 PVC pipe, and the loop was placed in a sand bed at the bottom of this saw cut. The cut was backfilled with stabilized base (two-sack portland-cement concrete) and the wear surface was patched with asphalt. While this approach used more capital funds, those capital funds were relatively more available than maintenance funds, and the resulting trend was an increasing percentage of functioning detectors.

Minimizing and Balancing Congestion

One of the recommendations made in the final report for the FHWA project *Signal Timing Under Saturated Conditions* was to devote the agency's best resources to its worst congestion problems. Leading practitioners, as reported in that work, recommended:

- Make use of every available feature, however esoteric, to improve green time efficiency in favor of the congested movements. This requirement might be seen as conflicting with a realistic understanding of maintenance limitations, but it need not be so. An agency can reconcile both these objectives by ensuring that the extreme operational efforts are spent wisely and not wasted on intersections that operate routinely and don't need such efforts.

- Devote the agency's best operational experts to the most congested problems, and be prepared to support their extended observation and experimentation in pursuit of a solution.

That research also suggested a clear focus on operational objectives as conditions moved into the congested regime. During uncongested conditions, the agency should try to provide smooth, predictable flow in accordance with the expectations of motorists. When congestion occurs, smooth flow is no longer possible, and the agency should switch to an objective of maximizing throughput. When demand increases to the point where queuing is inevitable, the agency should operate the signals to minimize the damage done by queue formation in an attempt to keep the problem from cascading throughout the network.

Following these approaches will, in many cases, provide a balanced approach to congestion that most motorists will see as (at least) equitable, consistent with their expectation of good basic service.

Smooth Flow

Many agencies now use advanced analysis tools to optimize and evaluate traffic signal operation. But the optimizations and the evaluation measures are surrogates for the "measurements" made by motorists. For example, motorists do not separately measure time delayed and time not delayed. They measure total travel time, and travel time reliability. ***All motorists apply a greater penalty to time delayed than time not delayed.*** But motorists also evaluate the cause of the delay when assessing blame, and delay caused by congestion is counted less against the agency than delay caused by red lights, especially when the other movements of the intersection are perceived to be underutilized. These situations are a leading cause of complaint calls.

The most successful agencies do not blindly optimize based on a narrowly cast objective function, such as delay, stops, or some combination thereof. Nor do they optimize solely on progression bandwidth. Rather, they provide a mix of operational objectives designed to be seen by motorists as fair and predictable.

For example, progression optimization might find a 10-second band through 20 intersections on a four-mile arterial system when using the optimally resonant, say, 90-second cycle. Such a progression band might be invisible to most motorists, who fall out of the band at the slightest disturbance in the flow, and are then repeatedly delayed until the next band catches up to them. Most motorists would perceive better basic service if this arterial was broken up into two or three segments, as long as the breaks in progression were clean (such that arriving traffic arrives on a long-standing red and not just as the signal is changing to yellow). With such breaks, much greater progression bandwidth might be possible. Instead of a large percentage of motorists being stopped unpredictably at many intersections, a higher percentage are stopped predictably at two locations (in addition to the first).

Another aspect of good progression design with respect to smooth flow is ensuring that the largest entering platoons see the greatest service by the progression band.

Optimization approaches that minimize delay and stops are often based on simplistic models of those stops. If the stops cannot be realistically modeled, optimizing on them is likely to lead to distorted results.

> ***By focusing on good basic service, an agency might forego an abundance of computer optimization and simulation, and the numerical data collection that such requires, in favor of simple and direct methods that ensure smooth flow, including time-space diagrams and on-street evaluation.***

Predictable and Consistent Response

Most highly responsive systems run into the problem of keeping the signals in useful coordination while the basic operation is being changed. Adaptive systems control this effect by limiting changes to small increments, though this may cause the operation to wander into timings that reduce the potential effectiveness of the timings. For example, some adaptive control systems allow small changes in the cycle length, even though a given system may provide good progression on only one or two cycle lengths within the acceptable range (as reported by Steven Shelby in his paper *Resonant Cycles in Traffic Signal Control*, Transportation Research Record #1925, 2005). Traditional traffic-responsive systems invoke often damaging transitions that are usually constrained to occur outside the peak period to avoid causing more problems than they solve. This is particularly true of transitions that dwell only, with shortway and subtraction transition methods causing less disruption. The transition problem suggests that fewer changes of the signal timing plan might be better in many situations.

Most agencies design only four or five signal timing plans for their arterial street signal systems. The range of conditions in any typical day, however, runs the entire gamut from a nearly empty network to peak conditions, and from a strong inbound bias to a strong outbound bias. When overlaying the range of possible solutions onto this wide range of conditions, most agencies divide that solution space into only four zones: Heavy inbound flow, heavy outbound flow, heavy balanced flow, and light flow. That suggests that signal timings must cover a broader range of conditions than the narrowly defined optimums many computer optimizers would imply.

Dr. Carroll Messer, author of the seminal PASSER II progression-based optimizer, observed during a past conference that broad progression-based solutions provided a *predictable* operation. Motorists routinely stop at the same intersections and therefore see the effect of the timing solution clearly. When that experience is consistent day in and day out, they can adjust their behavior (possibly) or their expectations to that reality. Systems that continuously change the motorist experience should be carefully selected when the need for that variability outweighs the expectation of a consistent motorist experience.

Signal Timing Versatility

Closely related to a predictable and consistent response is the need for versatility in signal timing. If the solution space is divided into only four zones, and if that solution space overlays a wide-ranging condition space, then each plan in the solution space must provide acceptable operation even at the boundaries of the condition to which it is applied. In highly dynamic cases, more than four solutions may be required, but that increased dynamic nature still will demand versatile signal timing solutions.

But most computer-based analysis of traffic signals applied by practitioners, both for optimization and simulation, considers the traffic demand as a steady state. Thus, signal timing approaches are evaluated by their performance during the narrowly unique conditions usually characterized by a 15-minute period rather than by the range of conditions over the hours of the daily schedule for which those timings will be used, and for the months and years over which that daily schedule will be used. The best agencies address this by one of two approaches:

- Implement a system capable of continuous optimization. That system may use normal traffic actuated features at the local controller, including gapping out, hold release, phase skipping, and volume-density features. Or it may use a more centralized or regionalized adaptive control based on network objectives. In all cases, the effectiveness of the system depends on detection.

- Design for versatility. The best agencies are able to understand signal timing effectiveness over a range of conditions by looking at the breadth of the solution rather than by seeking a narrow optimum. For example, a progression-based solution that works for heavy traffic will also work for light traffic. Even one car in the network will move unimpeded. This is a variation on Dr. Messer's observation, that such solutions a.) provide a predictable target into which motorists will fit their behavior, and b.) work over the broadest range of actual conditions. Agencies that are constrained by resources on a range of fronts can still maintain good basic service if they consider versatility in their methods and solutions.

Appendix I contains more development of the signal timing versatility concept.

IV. Interviews

To document current practice and to determine how superior agencies relate to the features of the above archetypes to provide basic service, the research team consulted two highly respected practitioners, one at the local level, and one at the state level. Because of the desire to hear their candid opinions and observations, the researchers have agreed not to report their identities directly.

Questions relevant to comprehensive management are listed below. These questions were a guideline for the interviewer, and were not intended to be asked directly to the respondent in all cases. The purpose of the interview was to answer these questions, but also to determine if the questions were valid or needed to be changed or supplemented.

1. What does your agency try to achieve?
2. What principles do you instill in your operations staff? Does complaint response get more priority than complaint prevention?
3. What do you think your motorists expect?
4. What do you think policy makers and elected officials expect?
5. Do you consider your operations staff more limited in numbers or in skills?
6. How do you respond to those limitations and expectations?
    a. Staff improvement?
    b. Staff supplementation (using consultants)?
    c. Productivity standards?
    d. Technological approaches (such as focus on simulation or avoidance of simulation)?
7. How do you know when you are doing a good job?
8. How much do you feel your operations staff are behind the curve rather than ahead of it?
9. How do you anticipate successful operations when setting design standards?
10. What constraints do you face on your ability to maintain your systems?
11. What do you do to keep your systems working as designed?
12. How do you define "operations?"
13. Do you think you are doing a good job?

Local Agency Traffic Signal Program Manager

The local agency manager expressed the reality that his staff are always so involved in dealing with problems that there is no time to fit their activities into an overall strategy. This is a common theme among agencies, and seems more related to perceptions of external expectations than to resource levels.

The manager prefers progression when possible, but recognizes that it is not always possible because of unresonant spacing and speed. He asks the following general questions when reviewing operation:

- Are we wasting time anywhere?
- Is delay distributed equitably?
- Is there green time serving no cars?
- Imbalanced queuing?
- Red when there's no competition?

Versatility is a high priority for the agency, but the manager would like a better means of evaluating how versatile is the operation. He has not considered a solution-space-based approach to organizing signal timings to ensure fundamental versatility.

For this agency, complaint response gets priority, of necessity, and because this is the measure most sensitive to elected officials and policy makers.

Motorists expect minimized travel time and the variability of travel time.

This agency manager believed that elected officials in his jurisdiction had different objectives than motorists. His perception was that they preferred not to think about their expectations, and only champion causes that affect how they are perceived by their constituents. Consequently they seemed generally unwilling to support programs that are not directly responsive to issues that are generating current public complaint.

The manager believed that increased staff would result in improved skills, because currently people are overloaded and stuck in reactive mode and don't have the opportunity to learn new skills, or to dig into situations so that they can really explore new alternatives. He also observed that those who improved their skills left the agency because of frustration related to too much of their time is being spent extinguishing brush fires.

The agency manager believed that they are doing a good job when they receive no complaints, but he also evaluated operation in terms of his own standards. But he believed that at the end of the day, there is often no good way to evaluate operation. He therefore advocates an automated means of observing traffic *performance* rather than *data*. He does not, however, have a plan for implementing such a capability, though he does advocate for the application of such technology initiatives as *IntelliDrive* to provide a link between vehicles and the infrastructure that would provide a description of time and distance trajectories that could be used to construct travel times.

He reports that his staff is behind the curve when addressing issues, with little hope of establishing proactive prevention.

The agency will not use advanced or esoteric controller features based on a realistic assessment of maintenance staff. They constrained actuated control by signal timing to prevent too much dependence on detection.

The manager sees imperfect understanding of true operation as the main constraint on agency effectiveness.

To ensure that signal timing stays as designed, signal timing settings in the controller are reviewed during preventive maintenance. The agency has a goal to drive streets systematically, but have never reached that goal, though they have noted improvement in the last several years. The time spent addressing complaints, which are managed by a separate part of the agency with extreme and vigorously enforced accountability rules, prevents systematic observation.

The manager described "operations" as the non-physical side of the road environment, including procedures for creating and modifying signal timings and signal configuration.

This agency reported a B- on the Signal Timing Report Card, but acknowledged that C+ was probably more appropriate. They reported a B- because they felt it was more politically acceptable, which the interviewee described as trying to convey that "we're not that good but we aren't as bad as most." He believed that the current self-assessment is not too bad, but still a little too focused on activities rather than results.

State Agency Traffic Signal Program Manager

The state agency requires using a business plan to define goals for activities—to optimize certain number of signals per year; deploying advanced systems; response time goals for maintenance.

For example, the agency defines a goal of 5% delay reduction for each optimization cycle, which is repeated every three years. This performance goal has had to be downgraded from 10% because repeated improvements proved elusive. (Author's note: This suggests that either traffic demand is increasing beyond what new signal timings can address, or signal timings are being redesigned more often than necessary.)

The agency holds monthly traffic meetings discussing traffic issues and to promote consistency. Resource limitations demand that complaint response receives the highest priority, and the manager perceives that the agency is always reactive.

The manager believes that motorists expect consistent, efficient, and safe operation. He prefers smooth progression along an arterial, not waiting when there is no demand on competing movements, good pedestrian accommodation, and phase protection only when needed. The agency maintains a strong concept of progression, preferring excellent progression in shorter segments rather than weak progression in longer segments. (Author's note: This requires departing from the results attained by software optimization tools currently in use by the agency.)

The manager perceives that elected officials do not want to hear from constituents about lack of efficiency and safety. He believes they are also concerned about the responsiveness of the agency.

The manager perceives that his operations staff is more limited in skills than in numbers.

The agency promotes staff improvement through in-house courses in control devices, which they also make available to consultants. For example, they have established a new mentoring process for construction inspection, where the designer and inspector are required to work together. The agency is trying to capture the skills and knowledge of staff nearing retirement. They also provide cross-training, and are trying to provide transparent access to their standard procedures.

The agency supplements staff using on-site consultants and consulting contracts to farm out a lot of design. They prefer to use consultants in a supporting role rather than a lead role, based on the belief that consultants are not good at signal timing, being too dependent on software and optimization, and lacking on street experience and insight in dealing with complaints.

The manager prefers going to the street to review operation rather than engaging in major retiming efforts or using advanced analysis tools such as micro-simulation.

The manager evaluates the program driven by complaints. The agency centrally manages signal timing, and believes that getting no citizen complaints means the agency is doing a good job. The manager maintains close contact with vocal citizens and believes in handling meaningful complaints using the most qualified staff.

To anticipate successful operation, the agency is doing several things, including replacing span wire with mast arms to reduce maintenance; installing LEDs; and using more video detection to reduce maintenance needs for in-pavement detection. They already limit detection schemes to simple approaches. The manager observed that video is a little more finicky and they are working to improve that. The agency's compliance with the Americans with Disabilities Act, which dictates pedestrian accommodation at intersections, is dictated by policy, where design is outstripping maintenance. The agency cannot afford the staff to be able to participate in the one-call program for preventing damage to their infrastructure by utility contractors, and this results in frequent damage to their system communications cable.

Because of resource limitations, the agency does not routinely review signals to make sure they are running the intended signal timings. They planned to perform complete optimization every three years, followed by annual reviews with a drive-through, but never had the resources to keep up with it.

The manager defined "operations" as timing, phasing, and coordination—anything outside of routine maintenance. He observed that designs and design changes are operations-driven though done by different people.

The agency scored a D on the original Signal Timing Report Card, but the manager thought questions were slanted to centralized operation and activities, and tainted by design (e.g. second-by-second communications). He worked to improve the questions in the self-assessment, but acknowledged that they also had problems—e.g. by evaluating based on ratios (number of signals per technician) instead of response time. The agency scored a grade of B the second time. The manager claimed they did not game the results, but are still pushing for more appropriate questions that focus on outcome more than output.

Observations

The results of these interviews reinforce the concept of basic service, but also reveal the difficulty even well-managed agencies have in articulating their objectives and designing their programs to proactively attain those objectives.

> *They also reveal that the initial Signal Timing Report Card focused too much on what agencies did rather than what results they attained.*

Both managers complained about the activity orientation of the Report Card. But neither agency was able to articulate what they did to measure their actual results.

Both agencies are driven by complaints in directing their activities, and both managers thought this reduced overall performance rather than increasing it. The agency managers did not attempt to express philosophies or judgments about the accuracy or representativeness of those complaints. It has been the author's observation that wealthy communities generate more complaints than poor communities, even when the service provided by the agency is better by any objective measure. One possible explanation is that citizens in locations that consistently provide less service lower their expectations and complaint less often. Thus, measuring overall agency effectiveness on the basis of complaint calls might, over time, trend towards a self-reinforcing cycle of lowered expectations and services. If so, then citizen complaints provide a strong accountability mechanism to encourage agencies to maintain strong service models, but when excessive consume the resources necessary to achieve those service models. The author has also observed that agencies have increased accountability for responsiveness to citizen complaints. Most agencies maintain a direct tracking system for complaints, expecting the response to complaints to be commensurate with the complaint itself (i.e., written complaints receive a written response, verbal complaints receive a verbal response, and so on). Both agencies devote their best communicators to responding to citizen complaints, primarily to help ensure that they neither come back nor escalate.

Another observation is that even though complaints are the only measure available to agencies, they don't often look for value in the complaint process. For example, if a high percentage of complaints result in a maintenance call, then that may signify an inadequate maintenance capability. If those maintenance-related complaints are commonly related to, say, detectors, that may indicate too much dependence on detectors, and a change in operational objectives and approach might reduce the dependence on detectors enough so that motorists won't notice when they require maintenance. An abundance of complaints regarding operational issues that have to be referred to maintenance forces may also indicate the inability of the signal system to allow the complaint responders to diagnose the complaint using the system. These trends could therefore lead to a better definition of needs and requirements in support of the next system upgrade. Usually, practitioners with experience develop a sense of the nature of complaints, but turnover eliminates this built-up understanding. Few agencies have systems that track complaints well enough to observe such trends without that experience.

Both agencies discussed training. The local agency manager complained that staff development was hampered by that staff being overburdened with complaint response activities. That agency does, however, provide opportunities for staff to become involved with national organizations and committees, and thus has maintained a high satisfaction at least for the more motivated staff members. The agency has not considered or articulated how it might make use of training resources already available, such as courses from the National Highway Institute, in support of staff development.

The state agency has a more explicit training program, and it is the author's general observation that this is true around the country. One way in which state agencies may assist local agencies with staff development is by sponsoring training programs and then inviting local agencies to participate by providing them grants to cover the expenses. The author noted a recent presentation of the NHI course *Traffic Signal Design and Operation* in Ohio that provided grants for local agencies to attend, and the course attendance included half a dozen local agencies in addition to state participants. State agencies, who generally have more resources to support such programs, may find that supporting local agency involvement helps the state agency meet its operational objectives as well.

V. Implementing a Basic Service Concept: The Traffic Signal Management Plan

Previously, this report presented the concept of good basic service as a reflection of agency archetypes that most effectively focused their resources, however plentiful or scarce, on their most important objectives. **The interviews revealed that even the best agencies have difficulty articulating and maintaining that focus.** This section will outline key strategies for doing so, and an outline of how to embody those strategies in a Traffic Signal Management Plan.

The following principles provide the basis for an agency to develop an objective-based traffic signal management program:

1. Clarity of objectives.

2. Attainable performance evaluation linked to those objectives.

3. Standards of performance (intended, for example, to score well on the Signal Timing Report Card) based on objectives rather than agency activities, especially arbitrary frequency of activities.

4. Resource requirements based on objectives rather than industry norms.

5. Clear and consistent communication with policy makers and elected officials.

6. Systems engineering in thought rather than merely in deed.

These strategies are discussed in the sections that follow.

Clarity of Objectives

As previously described, a good basic service concept is based on a simply stated objective. Drivers have simple overall expectations:

I want to drive to my destination at my desired speed with the minimum of attention. In the absence of achieving that goal, I want to be treated fairly and predictably so that I can plan my day with the minimum of uncertainty.

This leads to a simple objective for agencies to articulate:

We will do our best to avoid making drivers stop, but when we must make them stop, we will delay them as little as possible, within the context of safe operation.

As previously described, these objectives lead to a few high-level strategies for the agency to promote, including:

- Field infrastructure reliability.

- Signal timing that minimizes and balances congestion.

- Signal timing that promotes smooth flow. This objective is an amalgamation of more narrowly cast objectives, such as delay, stops, and so on, but those objectives are evaluated in light of the broader objective of smooth flow.

- Signal operation that corresponds to conditions predictably and consistently.

- Versatile signal timing that provides broad-banded solutions rather than being narrowly optimized for a specific condition. (See Appendix 1 for further discussion.)

Attainable Performance Measures and Standards of Performance

If the strategies listed above are relevant to achieving the goal of fulfilling motorist expectations, the performance measures should be related to those strategies. The interviews revealed that even the best agencies find it difficult to measure their performance (other than in terms of citizen complaint calls) and stay focused on their objectives. Too often, agencies focus on traffic performance objectives such as delay and stops that are based on aggregate measures that are not directly perceivable by motorists, and too often the achievement of good maintenance is hidden by the effects of traffic growth with respect to network capacity outside the control of traffic signal operations staff.

Thus, an effective operational plan will define how the agency has implemented and measured the progress of those strategies.

Field Infrastructure Reliability.

The broadest measure of the reliability of the traffic infrastructure is:

The percentage of time motorists see what traffic agencies intend for them to see.

There are two general approaches for accomplishing this objective that emerge from the discussion of agency archetypes: By design and by maintenance response. Maintenance response includes emergency and preventive maintenance, and agencies have difficulty establishing the latter based on their resources being consumed by the former.

Agencies that have successfully implemented preventive maintenance programs have often built new programs, with new technicians and new equipment, based on clear productivity objectives. For example, the City of Austin, Texas established a preventive maintenance program in the mid-1980's that resulted in the addition of two technicians, each with a small service van equipped with a short telescopic bucket. These technicians were entirely devoted to preventive maintenance, as a program for which funding was sought separately and distinctively by the agency staff from the local policy makers and elected officials. The program was based on the principle that the agency would not have to expand its maintenance force in future years to keep up with additional demand for emergency maintenance because of the effects of preventive maintenance.

Agency staff measured and reported the effectiveness of that program in terms directly linked to the objective of the program: How many traffic signals received preventive maintenance during the year? A performance expectation for the technicians was established at four intersections per shift for preventive maintenance, and the technicians worked a night shift to avoid congested periods. What the agency did not do was close the loop to analyze whether average emergency maintenance declined as a result of improved preventive maintenance, and as a result lost focus on the program after several years and folded the new technicians into the regular technician pool. As with the Report Card, performing the activity alone is not enough. With maintained focus, the agency would likely have been able to report a decrease in the number and severity of emergency maintenance calls.

The same is true on the system side for coordinated systems. In order to increase the percentage of time the motorists see what the agency intends them to see, the operation must actually be observed. The interviews revealed that agencies did not observe the operation of their signal systems routinely to make sure they were running as intended. To some extent, this forms one of the fundamental purposes of a traffic signal system, which is to monitor the local infrastructure and report faults when they occur. **Thus, the need for field observation can be minimized by a design solution rather than by merely increasing the maintenance function.** But requirements leading to such design must be clearly articulated.

Signal Timing

Measuring traffic performance can be related to two very different questions:

1. Has the signal operation degraded to the point where it needs to be revised?

2. Is the signal operation good?

The first question has usually been the result of an arbitrary time limit, and the Report Card evaluates effectiveness here on the basis of how often an agency revises its signal timings. But an activity-based solution is not necessarily an effective use of limited resources.

In an ongoing project considering this question, one result has been that signal timing designed to accommodate heavier traffic (short of congestion) will also accommodate lighter traffic, with little impact on travel time. That work defines the solution space for signal timing as ranging from nearly no traffic to the point of congestion in a typical day (which may be in the future). Considering that agencies usually provide effective signal timing with four or five coordination patterns for this comprehensive range of conditions, the research pointed out that trends over time could be accommodated by providing signal timing plans designed to cover the solution space at the outset. (See the discussion on versatility in Appendix 1.)

The thornier question is evaluating given signal timings in absolute terms, especially in congested networks. The interviews revealed a very simple metric for determining whether signal timings seemed reasonable and equitable to motorists: Citizen complaints. Thus, one means of measuring improvements in this objective is by summarizing citizen complaint activity. Raw numbers are probably not helpful, however, because notice in the press can cause a public reaction that would not otherwise have occurred. These anomalies should be measured and discounted. Most agencies now track citizen complaints. One of the interviewers, for example, reported that all complaints were processed by a central operator at the agency, and tracked to ensure that they are resolved timely. A reduction in the number of complaints per signal, or the failure of complaints to keep pace with growth, would indicate progress in this area. The prevention of a sudden and sustained increase in complaints would indicate that the agency has not slipped in this measure.

By measuring signal timing in this way, agencies can avoid the enormous task of attempting to measure traffic performance comprehensively. But for agencies that have the resources, there are ways to collect relatively simple performance data using travel time data collection, as suggested by some of the archetypes, using a pair of technicians (one for the morning period and one for the afternoon period) for each 200 or so arterial systems that are evaluated on an annual basis. These technicians can collect peak-period travel time data in half a dozen runs on an arterial, from Monday afternoon to Friday morning, keeping track of season-over-season or year-over-year trends.

Objective-Based Resource Allocation

With a basic service concept based on keeping traffic moving, many of the activities undertaken by agencies may be seen as tangential to their objectives. In developing a traffic signal management plan, each agency should consider how each activity actually contributes to achieving their objectives. A commitment to the operation on the street was reported as critical during the interviews, and the agency archetypes most committed to street performance can be seen to achieve the best results.

For example, an operations plan should include a description of how the agency will monitor signal operation to make sure it is operating as designed, and a description of how the agency will observe that traffic flow is still being accommodated. For agencies with severely limited resources, simple observation may be the only choice, but even that is often undone even by agencies that don't face such severe limitations. While agencies usually monitor their response to citizen complaints, they should also provide a mechanism for responding to complaints generated by their own staff. For example, the City of San Antonio published time-space diagrams for all coordinated arterial streets in a notebook, and provided those (with training in their use) to all traffic-department employees who spend their day on the street. Their instructions were to pay attention to flow anomalies, such as unexpected queuing or unexpected repeated stops. This is one simple and inexpensive mechanism for monitoring system operation. Approaches like this are particularly useful for state agencies that have widely dispersed small-scale systems that often don't provide communication to the traffic operations office.

The operations plan should also include a description of the mechanism by which these observations are followed up with review by operations staff. It is not the recommendation to impose a cumbersome tracking program, but observations from staff should be afforded at least as much accountability as citizen complaints, and this is often not the case.

The focus of the operations plan should be on results, not on activities. A system that provides skilled qualitative review of every coordinated system in the jurisdiction of the agency every year is preferable to a system that provides, say, microsimulation of only a few systems receiving public or political attention. The difference between these is the same as the difference between preventive and emergency maintenance. The resource-intensive analysis methods, including those that require detailed data collection and analysis, should be reserved for problems that cannot be solved by simple skilled observation.

But agencies should also include in their operations plan a commitment to the basic design of the signal timing for the system. A series of qualitative revisions to operation may be counterproductive if it undermines the basic operational concept. For example, a series of signal timing adjustments may ruin progression without the technician making the adjustment realizing it. All such adjustments should be vetted in light of the original intent of the operation, and if an adjustment is needed that can't be accommodated within that original concept, the agency should consider developing new timings.

Many such approaches are possible, but the important recommendation is that agencies don't ignore the responsibility to allocate resources based on operational objectives rather than merely on activities.

Clear Communication Up the Line

Most traffic signal practitioners prefer to work anonymously within their agencies. This attitude is a natural outgrowth of the negative experiences that often accompany such interaction. In many agencies, traffic signal staff are precluded from direct contact with elected officials, and frequently are put between the rock of citizen complaints and the hard place of agency upper management and public-relations staff that are incentivized to please politicians. This uncomfortable position often leads to a cynicism that can undermine the effectiveness of the agency more deeply than any budget shortfall. One common outcome is that responding to citizen complaints and requests for information from elected officials is often delegated to nonprofessional staff who are technically skilled but perhaps not able present the information consistently with agency goals and strategies.

An operations plan should include the description of a mechanism by which the progress of programs, including routine operational programs, is reported to agency management, policy makers, elected officials, and the public. The mechanism should allow both ad hoc and systematic reporting, and all the reporting should be based on operational objectives rather than on agency convenience.

These communications need not be excessively formal or highly produced. The City of Arlington, Texas developed a means of communicating to management in the form of a technical memorandum called an "Informal Report to Mayor and City Council." These should be short and descriptive, rigorously factual, and closely related to the interests of the motoring public. One of them, for example, reported on the results of an agency-wide signal retiming effort conducted to update timings five years after a signal system had been installed. As with all the informal reports, the organization of the report included a background, discussion, and conclusion. Improvements were based on actual measurement of travel times, using a floating car method. Analysis related these travel times to average speed, stops, delay, and emissions, but travel times received top billing. The report provided only a summary of the entire network, and was less than two pages. During each year, the agency developed dozens of these reports, and they were the basis for winning the ITE Urban Traffic Engineering Achievement Award.

The City of San Antonio was more limited in resources by far, but still established basic principles for communicating outside the operations staff. These were:

1. The story told to citizens is the most important story in the agency. If it is ineffective in achieving citizen support and satisfaction, it will also not achieve policy and political satisfaction. Consistent with this importance, all citizen complaints concerning signal operation were not delegated, but handled directly by the professional in charge of the program. This had the dual benefit of telling the story most effectively, and making sure that the creator of the policy learned how to express it most effectively.

2. Requests for information from top management and elected officials must be responded to meaningfully within seven calendar days. This included those requests that required data collection, such as signal warrant studies.

3. Information presented to upper management would be presented with a high standard of professionalism, empowering and enabling those upper managers to represent that information to policy makers and elected officials more effectively. For example, the before-and-after travel-time studies associated with new traffic signal timings were summarized in a professionally produced report that extended the memorandum concept mentioned above to a report that could be distributed to the public.

An operations plan should include principles for communicating outside the agency, with special emphasis given to reporting successful completion of routine and special projects, and to making it the direct responsibility of the person in charge of the program to articulate it to the public. Those principles should be directly related to the objectives of the program. For example, a new signal system might be built to improve the management of field infrastructure, in which case the report should be related to that objective, e.g. the number of citizen complaints that were responded to without a field visit because of the system, the number of faults that were discovered and corrected before the public complained, the increased number of intersections updated with new and improved timings as a result of more efficient database management, and so on. Signal timing projects, on the other hand, should show the benefits of new timings in terms most relevant to motorists: Minimize travel time and travel time unreliability, which is usually interpreted as unpredictable and inequitable operation by motorists.

Meaningful Systems Engineering

An operations plan should include a requirement that all projects follow a meaningful systems engineering approach. This requirement should apply to all activities defined as projects, with an objective, a strategy, a tactical implementation, and a result.

This is particularly true for the replacement of signal systems, which are the most expensive and perceived to be the least successful of traffic signal management projects.

Most signal systems are justified based on traffic performance and safety. *Yet these improvements result from signal timing improvements, not signal systems, and can be achieved through such approaches as time-based coordination.*

Most agencies contemplating a signal system, however, have real needs not served by time-based coordination. Those needs can be summarized simply: Manage field infrastructure.

Systems engineering is a process by which real needs are identified first, and then the project is formulated and tested to confirm that those needs are addressed. As an established branch of systems development, systems engineering is well documented in wide-ranging literature (in addition to being a requirement for Federally funded projects, see here: http://www.ops.fhwa.dot.gov/int_its_deployment/sys_eng.htm) and generally follows this process:

- Document what the agency does using the proposed system. (This is not what the system does, which will be developed later, but what the people do that the system must support. As such, it doesn't describe the system at all; it describes the agency.) These activities are then broken down into a list of needs, called user needs. The documentation of this step is called a *Concept of Operations.*

- For each need, define a requirement or series of requirements that the system must fulfill in order to address that need. These requirements form the *Requirements Document* and show explicitly how the requirements trace back, point by point, to the user needs.

- Design the system such that all the requirements are fulfilled. This is the first step where the design of the system is contemplated, and this is the first description of the system itself. The design includes the normal technology analysis, design documents, and ultimately plans and specifications. Every design feature must be justified by playing a role in fulfilling at least one requirement, and every requirement must be fulfilled. One of the design documents is the *Requirements Traceability Matrix,* which explicitly documents how the design fulfills the requirements.

- Test system components against the design. This is called *Verification.*

- Test the system's ability to support the activities defined in the *Concept of Operations.* This is called *Validation.*

With that outline, a signal system project Concept of Operations will include a plain description of what operations staff will do. For example, when a citizen calls the practitioner with a complaint about the operation of a signal, the practitioner needs to be able to observe the operation of that signal in real time to evaluate the complaint. Many causes of the complaint can be observed by watching the signal operate through the system. Thus, the agency's activity is to respond to a citizen complaint by first observing the operation of the signal in real time, checking that:

- The signal is not on conflict flash
- The signal is in coordination
- The pattern is correct
- The controller time of day is correct
- The timing has not been modified
- The detectors are cycling normally
- The signal is cycling normally

Further, the practitioner may benefit from seeing the traffic demand at the intersection using some form of traffic monitoring.

Based on what the observation reveals, the practitioner may dismiss the complaint, determine that a malfunction has occurred and refer it to maintenance technicians (or repair it through the system), or determine that the signal timing needs closer observation for possible revision.

The forgoing description of the agency's activity is the critical basis for determining the need for a traffic signal system. They provide a clear basis for requirements that will be used to evaluate the design of the system, such as the requirement to provide a real-time display, the requirement to include detector actuations in that display, the requirement to build a communications network that can keep up with that real-time display, and so on. Those activities should therefore be the subject of a proper Concept of Operations. The requirement, for example, that the system software include a real-time display can be traced directly back to this activity.

Most agency-developed systems engineering approaches focus on the system at the outset, rather than on the activities of the agency, with the result that the system eventually does not fully support those activities, and the project fails to meet expectations. This is especially true if the system was sold to policy makers on the basis of traffic improvements, which may be elusive, especially if traffic demand is increasing fast enough to mask any resulting improvements in signal timings.

Some agencies now require a business-plan approach for all their activities, and the systems engineering documentation will provide the necessary rigor to describe those activities, their objectives, and how those objectives will be measured.

Meaningful systems engineering is crucial to incorporating new technologies, as the above examples illustrate. An operations plan should provide an outline for a systems engineering process to be implemented for each routine and special project undertaken by the agency, following established approaches. These steps need not be excessively complicated. For example, many projects undertaken to support a particular activity might be described in a page or two. Even routine system projects can use a model concept of operations developed by the agency or within the industry, keying on the activities needed by that agency.

Sample Outline for a Traffic Signal Management Plan

Chapter 1. Objectives and Requirements.

In this chapter, define in concrete terms the operational objectives of the agency. While the definition should be concrete, it does not have to be specific. For example:

We will do our best to avoid making drivers stop, but when we must make them stop, we will delay them as little as possible, within the context of safe operation.

This mission statement is not specific, but it is concrete, and it can be used as the basis for establishing high-level requirements for the agency. Other objectives are possible, given local priorities, and based on the expectations of motorists in the region. The objective should avoid, however, specific performance standards, because traffic signals do not create capacity, and any given performance standard will fail when demand exceeds capacity.

Based on the objectives, the agency should establish high-level requirements that lead to strategies implemented by the agency. These strategies must be linked to the objective. These requirements may include:

- Reliability standards, based on the percentage of time the motorists see what the agency intends for them to see.

- Signal timing standards including:
    - Minimizing congestion and managing unavoidable queuing.
    - Promoting smooth flow
    - Appropriate for conditions
    - Versatile to provide acceptable operation across the range of conditions

Note that these requirements do not impose either a particular design or even a given level of resources. In the face of limited resources, creative design may fulfill these requirements, as described throughout this report. Some examples:

- Improving reliability by improving design and initial construction quality, by minimizing the dependence on features and infrastructure (particularly detection) and by using system approaches that do not depend on fragile communications infrastructure.

- Improving versatility by designing signal timings that are purposely broad-banded, such as progression-based solutions, rather than narrowly optimal as might be suggested by some optimization programs. Or, alternatively, strongly funded agencies might fulfill this requirement by implementing systems that continuously respond to changing traffic conditions based on measured inputs. Both of these fulfill this high-level requirement, but they do not both make the same resource demands.

This chapter does not need to include these strategies, but should establish the standard by which these strategies will be evaluated. These standards become the agency's definition of *good, basic service*.

Chapter 2. Responsiveness to Citizens, Media, Policy Makers, and Elected Officials

This chapter should lay out the standards defining the agency's responsiveness to motorists and other users of traffic signals. Examples of these standards include:

- All citizen complaints will be responded to in kind. Telephone complaints will receive a telephone response, written complaints a written response, and so on.

- All citizen complaints will be recorded by the person receiving the call, and responded to within a certain time period to be determined by the agency.

- All complaints should be checked to rule out malfunctions first. This strategy provides a range of requirements for signal systems and their ability to provide fault and real-time monitoring.

- Operational complaints that are not malfunctions should be addressed by the person who is responsible for the design. Explaining a design should never be delegated.

- All requests requiring data collection and studies should be addressed within a given time period defined by the agency.

- All complaints and requests should be tracked. The complexity of the tracking method is a strategy unique to each agency, but at the very least, the tracking method should ensure and be able to demonstrate that all complaints and requests are completely and expeditiously addressed. Most agencies already track requests coming from policy makers and elected officials, but the agency should also find ways to track common citizen complaints.

- The agency should define how the traffic signal operations staff should or should not communicate with the media. Many agencies already have such policies in place, but within what is allowed by those policies, agencies should ensure that the highest priority is given to the story told to the public. That usually means providing the agency's most articulate, positive, and knowledgeable explainer to respond to media requests.

- The agency should define standards for communicating program and project results to policy makers, elected officials, and the public. Those standards should include the format for such communications, an explanation of the objective of the program or project, the approach taken to achieve the objective, and the results in terms relevant to the objective. Agencies should describe here which programs and projects will be communicated this way, but it should at least include all programs and projects that receive specific funding, either within the budget or from outside sources.

Chapter 3. Maintenance Strategies to Achieve Objectives

This is where the agency lays out its maintenance approach. This chapter should include a clear explanation of how the agency will respond to limitations and constraints in maintenance resources, including limitations in staff numbers, training levels, and equipment. These strategies may include, for example, how an operational approach that does not depend on extensive detection minimizes the need for a large and difficult to maintain detection infrastructure. They should include how consistent design standards will be used to prevent sub-standard installations that require high maintenance levels. They should include how the use of features and technology within the signal equipment will be kept within the knowledge, skills, and abilities that can be expected for the job descriptions and remuneration for maintenance technicians.

Chapter 4. Operations Strategies to Achieve Objectives

In this chapter, the agency will describe the following:

- The operational approach. Examples include maximizing progression, eliminating onset of yellow during platoon arrivals, pedestrian accommodation, clearance interval approach, basic detection strategy (consistent with the agency's resources), signalized left turn warrants, allowable phase sequence approaches, and so on.

- Evaluation. How will the agency ensure that the operation on the street meets objectives, including operating according to design, providing smooth flow, providing sufficient versatility, and so on. The evaluation approaches can be quantitative (e.g. travel-time studies) or qualitative (field observation), but they must be directly linked to the objective. If the objective is to keep traffic moving, then the evaluation must assess how well traffic moves. If the objective is to maximize throughput through a bottleneck, then the evaluation should measure throughput. If the objective is to manage queuing at a bottleneck, then the evaluation should directly evaluate queuing. If the objective is to improve safety at a location, then the evaluation should include tracking crashes.

Chapter 5. Design Strategies to Achieve Objectives

Most agencies already have documents guiding infrastructure design requirements. But those documents do not usually set quality standards as they affect operations and maintenance, and this chapter provides an opportunity for agencies to do so. This section should apply consistent standards for traffic signal design, including street furniture, pedestrian accommodation, numbers and placement of signal heads, and so on. Not all agency operations staff are responsible for these design standards, in which case they should be written down here with an analysis of how the standard does or does not impose limitations on operations and maintenance. This provides the operations practitioner with an understanding of how design affects operations.

A most important section in this chapter concerns the implementation of automated systems, either for traffic signal operation or for management (including, for example, a citizen complaint tracking system). This chapter should lay out a meaningful systems engineering approach, including how each project (internal or external) will progress from a clear statement of needs through a demonstration of how the finished project supports those needs. The process should include at least the following steps:

- The Concept of Operations, including how an agency will use the system. The agency should write this document internally, or be the primary audience for a document written by consultants. In all cases, it should focus on what they will do, not on what features they want.

- Requirements. Each activity described in the Concept of Operations imposes one or more requirements. For example, an activity that includes an operator reviewing the real-time operation of the system leads to requirements that establish that real-time capability, and a display capability that shows the operator what they need to see to complete that activity. Those requirements become the standard for design.

- Verification. In this step, the designer demonstrates how each element of the design fulfills the requirements. This term also describes how the system will be tested to prove that it conforms to the design.

- Validation. Here the implementer demonstrates that the system indeed supports the activities described in the Concept of Operations.

Appendix 1. Signal Timing Versatility Concepts

Versatility is one of the more critical points in objectives for signal timing programs, but it receives little attention or consideration by many practitioners. The typical current approach to performing signal timing is as follows:

1. Collect physical intersection and network data and enter into signal timing optimization software
2. Collect 15-minute turning-movement counts at all intersections
3. Collect road-tube counts at all intersections
4. Review data to determine when traffic patterns change significantly, with "significantly" being defined to fit a pre-determined number of signal timing plans
5. Select a representative 15-minute period for each period of unique traffic pattern
6. Convert the 15-minute volumes to hourly volumes, and enter them into signal timing optimization software
7. Review the output of the optimization software and make adjustments, usually using a time-space diagram to represent the coordination pattern
8. Implement the timings on the ground
9. Fine-tune the timings on the basis of field observation

Several of these steps are quite expensive, particularly including the collection of 15-minute turning-movement counts at all intersections. This data-driven process implies that traffic volumes are more important than traffic performance, given the resources required to collect them. A key weakness of this approach is that volume data is of limited value to the practitioner, and often too expensive to collect routinely.

Practitioners use volumes primarily because optimization software requires it. And the software requires it because we have no easy means of directly measuring what is really important, and that is whether the timings are serving the representative desires of the users of the network. These desires represent aspects of the overall motorist objective for smooth, predictable flow, and we characterize them generally based on some combination of the following objective functions:

- Minimize delay
- Minimize stops
- Minimize fuel consumption
- Minimize emissions
- Maximize progression
- Maximize throughput
- Manage queues
- Minimize travel time (or travel time unreliability)

- Maximize right-of-way for non-automotive travelers, including pedestrians, bicycles, and transit riders

Modeling these objectives, especially considering that all the objectives are important at some level in all situations, is difficult even for microscopic simulation software. Most signal optimization software chooses some simple combination of a few of these objectives, and even these are quantified by simplistic models. The models themselves have been designed to use the only traffic data that is readily available, and that is traffic volume.

The traditional approach therefore starts with an expensive data collection process, which is required by a simplistic model that considers too few objectives, to produce signal timings optimized for a single 15-minute period which may or may not be usefully representative of other periods, especially over the coming months and years.

Skilled practitioners usually then are compelled to spend considerable time making field adjustments of the resulting timings, based on their experience-driven and implicit understanding of meeting the needs and expectations of as many of their users as possible. This process is also resource-intensive, and not all agencies have access to such skills.

Finally, the data on which the system is designed changes constantly. Timings narrowly optimized for several 15-minute periods may not provide the versatility needed to remain reasonably close to an optimum throughout any given day, or over time as daily, weekly, seasonal, and long-term trends affect demand. This leads to the usual recommendation to perform the above process again in an arbitrarily short time, such as every two years.

Tools are not available to evaluate signal operation on the basis of better-defined objectives and optimization processes that use measured performance rather than modeled performance.

Rather than the typical steps listed above, a better approach that is more sensitive to available resources and more versatile, might be the following:

1. Identify intersections where minor movements are controlled by non-vehicular factors. There is no need to collect data if the minor movement green time will be controlled by pedestrian crossing times, or if the traffic is obviously so light that any reasonable minimum green will serve it. In many cases, timings on these approaches can be set arbitrarily, with local actuation shortening the green time as allowed by demand.

2. Collect turning movements only at intersections where minor movement green times cannot be estimated.

3. Identify network topology, primarily the spacing between signals and the prevailing speed of traffic.

4. Choose the range of cycles lengths, based on coordination timings that resonate between signal spacing and prevailing speed.

5. Choose offsets (and, where desired, phase sequence), based on progression and directional split, for a comprehensive variety of directional splits. Directional splits rarely exceed 70% in the peak direction, and five plans should be sufficient to represent all directional splits from 70% inbound to 70% outbound on the most extreme arterial street. Most streets would need only three levels of directional bias.

6. Determine green splits at intersections where minor movements are not dictated by non-vehicular factors. First, identify the directional splits in the turning-movement data (by

inspection), and choose 15-minute periods within those directional split periods to characterize minor-movement demand. Then, determine green times equitably.

7. Implement and fine-tune.

Because the cycle length, offset, phase sequence, and main-street green time are largely determined by the geometry of the network, using volume-based optimization tools often has little additional to offer. Finding an appropriate progression solution may be difficult in more complicated networks, but even "guess" volumes can often find those solutions as accurately as measured volumes, especially considering the potential inaccuracy of those measurements. Often, optimization processes work to levels of precision far in excess of needed or supported accuracy, which is an error most engineers are trained to avoid.

The figure below represents the concept of versatility.

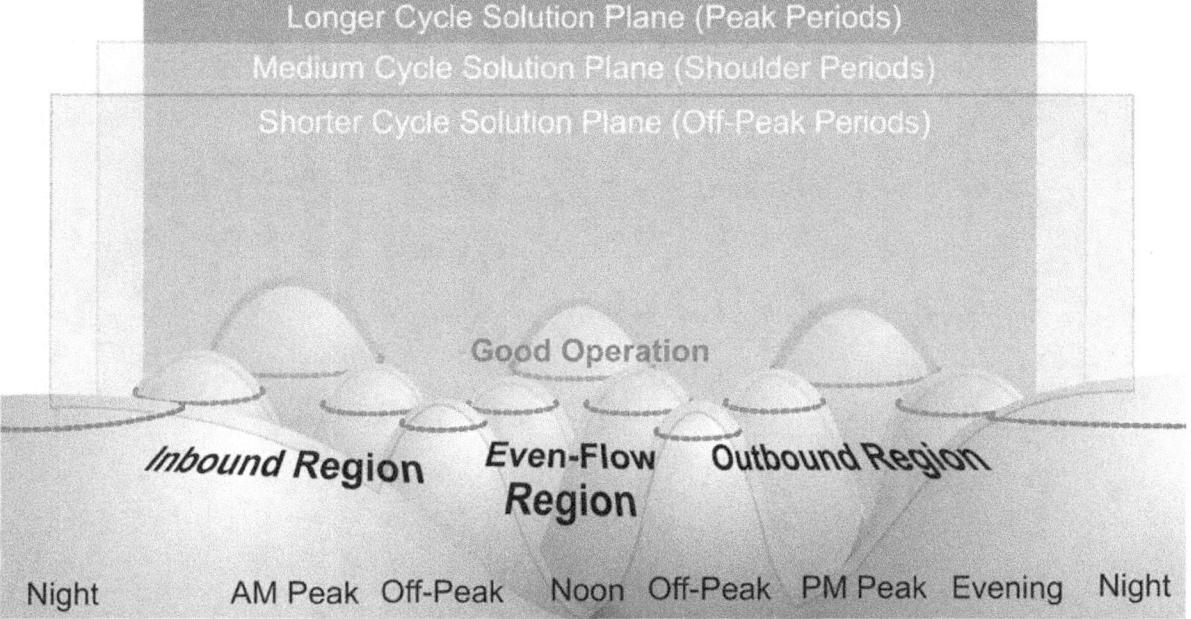

Good operation is represented by the peaks in the solution terrain, and good operation can be maintained by jumping from peak to peak. The current methods used by many agencies attempts to identify a few peaks to broadly represent the terrain, rather than to start with the whole solution terrain and represent it with versatile timing patterns. Using this approach, the most complicated arterial street may be timed with a dozen plans, none of which requires data at most intersections and only a few of which require significant data or design resources. Most typical arterials would need less than half that many plans, especially if the signal timings are designed for versatility.

Such an approach would provide *good basic operation*, saving expert resources for those conditions and geometries that demand more effort, such as congested conditions.

It should be pointed out that a versatile set of signal timing plans designed to broadly accommodate the solution terrain can support the implementation of traffic-responsive operation far more easily, and are also much more resistant to degradation over time.

www.ingramcontent.com/pod-product-compliance
Lightning Source LLC
Chambersburg PA
CBHW081908170526
45167CB00007B/3207